Science Inquiry

Habitats

by Joe Baron

Science Inquiry

Science in a Snap!

Explore Activity

Investigate Habitats 6

▶ **Question:** Which plants and animals live on land and which live in water?

Directed Inquiry

Investigate How Desert Plants Survive 10

▶ **Question:** How can the waxy covering of a leaf help a plant survive in a dry desert?

Think Like a Scientist **Math in Science** 14

Tables and Graphs

Science in a Snap!

What's That Habitat?

Draw an animal in its habitat. Tape the drawing to your partner's back. Your partner should tape a drawing on your back, too. Don't peek! Ask each other yes-or-no questions until you guess what animals and habitats are in the drawings. What might your animal need to survive in its habitat?

Droopy Celery

Put a wilted stalk of celery in a cup of water that has blue food coloring in it. Observe the celery the next day. What happens to the celery? What can you infer about what happened to the celery in the water?

Lights, Plants, Action

Put two sprigs of elodea in a clear bottle ¾ filled with water. Place the bottle in a sunny place or by a bright light. Use a hand lens to observe the leaves of the plant. What do you see?

Investigate Habitats

Question Which plants and animals live on land and which live in water?

Science Process Vocabulary

observe verb

When you **observe,** you use your senses to learn about an object or event.

water lily

duckweed

cactus

Plants

swift fox

prairie dog

angelfish

Animals

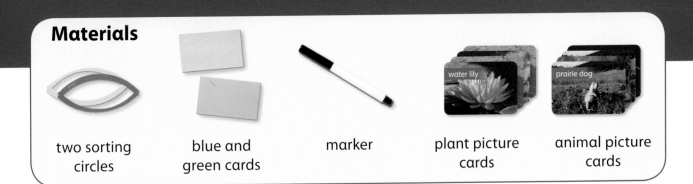

Materials

two sorting circles	blue and green cards	marker	plant picture cards	animal picture cards

What to Do

1 Unfold your sorting circles.

2 Make 2 habitat cards. Write **water** on 1 card and **land** on the other.

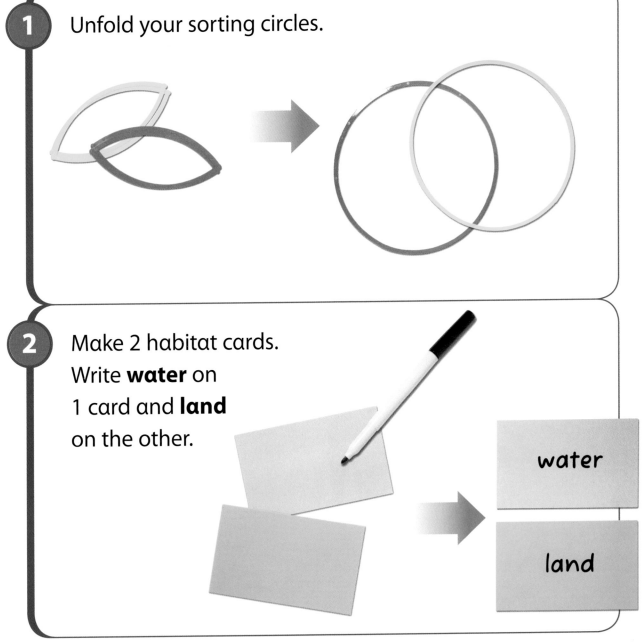

water

land

3 Place one habitat card near each sorting circle.

water

land

4 **Observe** each picture card.
Sort the plants and animals by where they live.

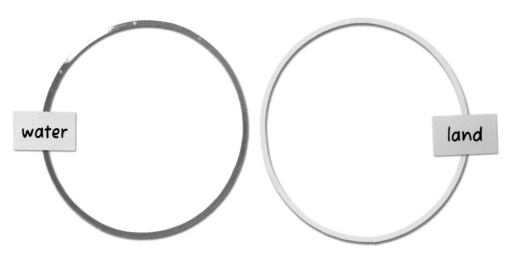

cactus

angelfish

swift fox

duckweed

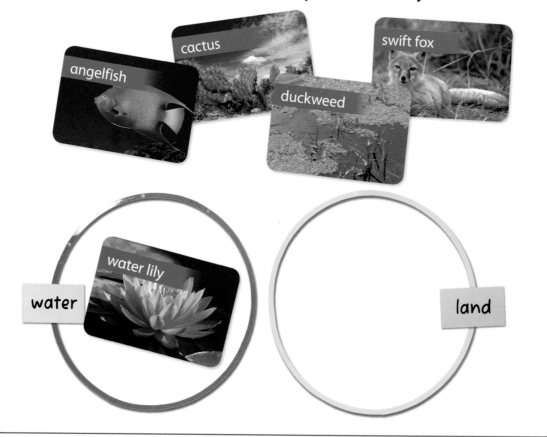

water lily

water

land

Record

Write or draw in your science notebook.
Use a table like this one.

Water	Land
water lily	

Share Results

1. Tell what you did.

> I sorted _____ and _____.

2. Tell where each plant and animal lives.

> The _____ lives _____.

Investigate How Desert Plants Survive

Question How can the waxy covering of a leaf help a plant survive in a dry desert?

Science Process Vocabulary

model noun

You can make a **model** to show how something works.

predict verb

When you **predict**, you tell what you think will happen.

I predict that spraying the paper will make it look darker.

Materials

green construction paper   scissors   spray bottle with water   waxed paper

What to Do

1 Draw two leaf shapes. Cut out the leaf shapes. These are **models** of leaves.

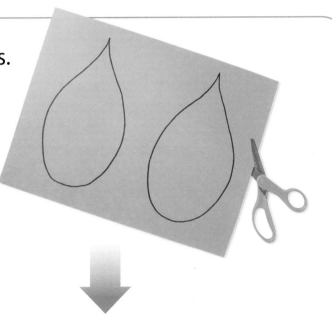

2 Spray both leaf models with the same amount of water.

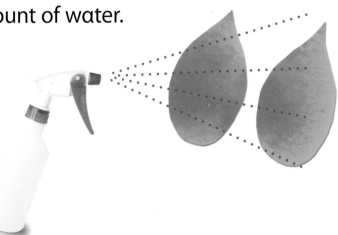

3 Fold the waxed paper in half. Place one leaf model inside.

4 Put both leaf models in a sunny place. **Predict** which one will hold water longer. Write in your science notebook.

5 **Observe** the leaf models for 3 hours. How do they look? How do they feel?

Record

Write or draw in your science notebook.
Use a table like this one.

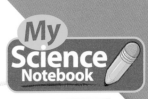

Leaf Model Cover	At Start	1 Hour	2 Hours	3 Hours
Waxed paper				
No cover				

Explain and Conclude

1. Which leaf **model** held water longer?
   Does this support your **prediction?**

2. Explain how leaves with a waxy covering might
   help a plant survive in a dry desert.

Think of Another Question

What else would you like to find out
about how plants can survive in their
habitats? What could you do to answer
this new question?

Math in Science

Tables and Graphs

Sometimes scientists organize data in a table. They can use the data to make a graph. This makes it easier to think about the data they collect when they investigate.

Kara wanted to find out what kinds of animals live in one land habitat. She counted the number of each kind of animal. Then she organized the data in a table.

Table

Animal	Cat	Dog	Bird	Squirrel
Number of each	2	3	2	1

Kara used the data to make a graph. First, she drew a column for each kind of animal. She wrote the names of the animals under the columns.

Then Kara made 4 rows. She put numbers by the rows. The numbers tell how many animals she saw. Kara wrote a title for her graph.

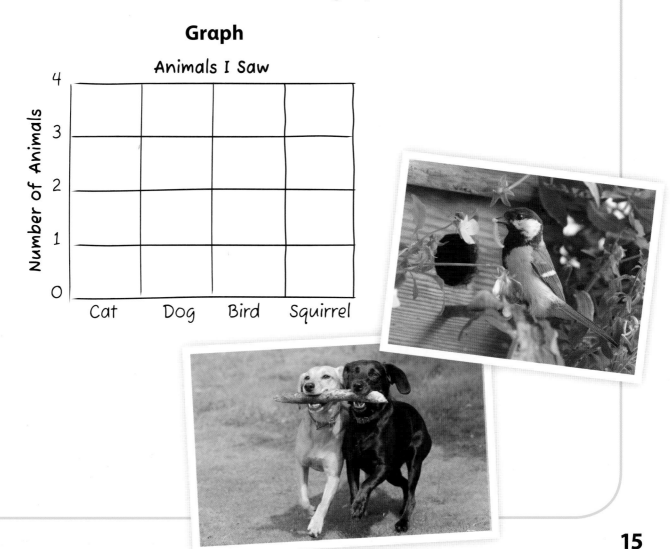

Graph

Animals I Saw

	Cat	Dog	Bird	Squirrel

Number of Animals: 0, 1, 2, 3, 4

Kara colored the graph to show how many animals she saw. She saw 2 cats, so she colored up to the line labeled **2**. She colored the number of spaces for dogs, birds, and squirrels.

Now Kara can quickly compare the number of each kind of animal she saw.

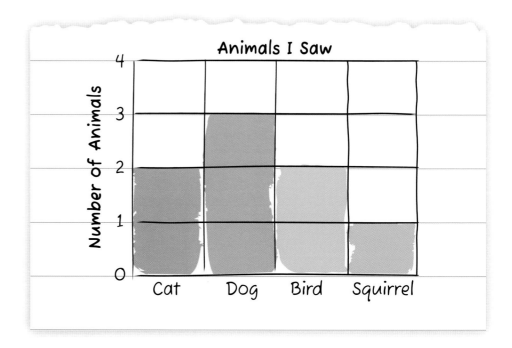

► **What Did You Find Out?**

1. What does the graph show?
2. How many birds did Kara see?
3. Which group had the most animals?

Make and Use a Table and a Graph

1. Look for land animals near your home. Gather data. Make a table.

   - Choose 4 kinds of animals.

   - Write how many animals of each kind you saw.

2. Use your data to make a graph.

   - Write the group names on the bottom.

   - Write the numbers on the side.

   - Write a title.

   - Color in boxes to show the number of animals in each group.

3. How does the graph help you to understand the data better?

Investigate a Mealworm Habitat

Question What happens if mealworms are given more than one type of food?

Science Process Vocabulary

predict verb

When you tell how you think an investigation will turn out, you **predict.** Sometimes you describe your prediction by saying "If ____ , then ____ ."

> If I water a plant, **then** it will grow.

fair test

An investigation is a **fair test** if you change only one thing and keep everything else the same.

Materials

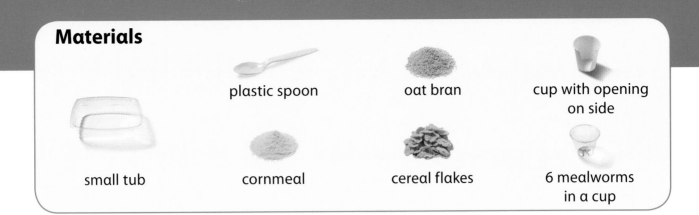

plastic spoon

oat bran

cup with opening on side

small tub

cornmeal

cereal flakes

6 mealworms in a cup

Do a Fair Test

Write your plan in your science notebook.

Make a Prediction

Mealworms eat food that comes from plants. In this investigation, you will give cornmeal to mealworms. You will choose a second food for the mealworms. Will more mealworms move to one of the foods? Write your **prediction.**

Plan a Fair Test

What one thing will you change?
What will you observe or measure?
What will you keep the same?

What to Do

1 Put 1 spoonful of cornmeal in one corner of the tub. Put 1 spoonful of another type of food in the next corner of the tub.

2 Place the small cup in the middle of the tub. Be sure the open side faces the food. Use the plastic spoon to place 6 mealworms in the middle of the cup.

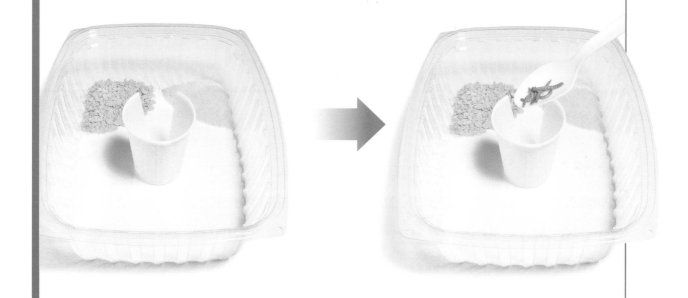

3 After 30 minutes, **observe** where the mealworms are. Record your observations in your science notebook.

4 Observe where the mealworms are every day for 2 days. Record your observations in your science notebook.

Record

Write in your science notebook. Use a table like this one.

Number of Mealworms		
	Cornmeal	My Choice: _____
Number of mealworms on day 1		
Number of mealworms on day 2		

Make a graph of the number of mealworms on Day 2.

Mealworms and Food, Day 2

Number of Mealworms

6
5
4
3
2
1
0

Cornmeal   My Choice: _____

Explain and Conclude

1. Did more mealworms move to one of the foods? How can you tell from your graph?

2. Do the results support your **prediction?** Explain.

Think of Another Question

What else would you like to find out about habitats of mealworms?

21

Do Your Own Investigation

Question

Choose a question or make up one of your own to do your investigation.

- What happens if you plant a water plant on dry land?
- What happens if some bean seeds get water and others do not?
- What happens if you give mealworms a choice of dark or light habitats?

Science Process Vocabulary

data noun

You collect **data** when you gather information in an investigation.

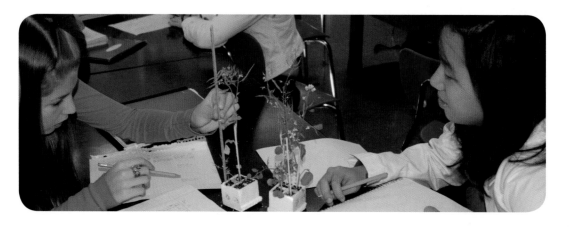

Open Inquiry Checklist

Here is a checklist you can use when you investigate.

- ☐ Choose a **question** or make up one of your own.

- ☐ Gather the materials you will use.

- ☐ Tell what you **predict.**

- ☐ Plan a **fair test.**

- ☐ Make a **plan** for your investigation.

- ☐ Carry out your **plan.**

- ☐ Collect and record **data.** Look for **patterns** in your data.

- ☐ Explain and **share** your results.

- ☐ Tell what you **conclude.**

- ☐ Think of another question.

How Scientists Work

Solving Problems Together

Many prairies once grew in Illinois. Now farms, roads, and cities cover the land. Most prairies are gone. Many animals that depend on prairie plants cannot survive.

Change in Illinois Prairies

Prairies, 1820
● Prairies, 2009

Some scientists want to save more prairie habitats in Illinois. These scientists are collecting seeds to grow new prairie plants. They search for the best places to collect seeds and where to plant them. They tell other scientists what they find out.

Scientists know that fire kills weeds, shrubs, and trees that crowd out the prairie plants. Then prairie plants survive and grow better after the fire.

Scientists work as a team to study the best times of the year to burn prairies. They share what they learn so more prairie habitats can be saved.

▶ What Did You Find Out?

1. How are scientists working together to save prairies?
2. How do you think burning the prairie affects people and the environment nearby?

◄ Prairie plants survive fires because they have deep roots and underground buds.

 ## Solve a Problem with a Team

1. Set a goal to recycle 100 sheets of paper.

   • First, collect paper on your own.

   • Then, collect paper as a team. Put your collected paper together.

2. Tell one way that working on a team helped you solve the problem.

3. Tell how recycling paper might affect other people and the environment.

Featured Photos

NF
Nature

Cover: gilded flicker in saguaro cactus nest

Title page: wood frog beside mushroom

page 9: eastern lubber grasshopper

page 10: potted agave plant

page 13: potted agave plant

page 23: young bean plant

page 27: skipper butterfly drinking nectar from a purple coneflower

inside back cover: bufflehead duck with chicks in tree trunk nest

ACKNOWLEDGMENTS

Grateful acknowledgment is given to the authors, artists, photographers, museums, publishers, and agents for permission to reprint copyrighted material. Every effort has been made to secure the appropriate permission. If any omissions have been made or if corrections are required, please contact the Publisher.

PHOTOGRAPHIC CREDITS:

set-up photography: Andrew Northrup; stock photography: Cover (bg) John Foxx Images/Imagestate; Title, Inside Back Cover (bg) Michael Quinton/ Minden Pictures/National Geographic Image Collection; 2 (t), 20 (t), 23 (t) Corey Hochachka/Design Pics/Corbis; 6 (t) Gary Bell/zefa/Corbis; 6 (tl), 7 (tcr), 8 (bl) 2265524729/Shutterstock; 6 (br), 8 (cr) Juniors Bildarchiv/Alamy Images; 6 (cl), 8 (cr) Roger Eritja/Alamy Images; 6 (cr), 7 (tr) Raymond Gehman/National Geographic Image Collection; 6 (bl), 8 (cl) EuToch/Shutterstock; 6 (br), 8 (cl) Digital Vision/Getty Images; 9 (b) Joseph H. Bailey/National Geographic Image Collection; 10 (t) Panoramic Images/Getty Images; 13 (b) Richard Griffin/Shutterstock, (bl) Marcel Jancovic/ Shutterstock; 14 (t) Bob Blanchard/Shutterstock, (br) Igor Burchenkov/Shutterstock; 15 (t) iofoto/Shutterstock, (c) Sharif El-Hamalawi/iStockphoto, (b) Jean Frooms/Shutterstock; 17 (l) IPS Co., Ltd./Beateworks/Corbis, (r) Simon Hadley/Alamy Images; 18 (t) Pakhnyushcha/ Shutterstock, (b) Liz Garza Williams; 22 (t) Dorling Kindersley/Getty Images, (b) MShieldsPhotos/Alamy Images; 23 (b) PhotoDisc/Getty Images; 24-25 (b) Leigh Prather/Shutterstock; 25 (c) M. Williams Woodbridge/National Geographic Image Collection; 26 Larry Miller/Photo Researchers, Inc.; 27 (t) Jim Richardson/ National Geographic Image Collection, (b) Darlyne A. Murawski/National Geographic Image Collection.

ILLUSTRATIONS:

pages 4 and 9: Amy Loeffler; page 24: Jeannie Barnes. "Changes in Illinois Prairies." Land Cover of Illinois in Early 1800's [shapefile], Champaign: IL Natural History Survey, 2002 and IL Natural Areas Inventory [shapefile], Springfield: IL Natural Heritage Database, IL Dept. of Natural Resources, 2009. Using: ESRI ArcView GIS [GIS software]. Version 3.3.

PUBLISHED BY NATIONAL GEOGRAPHIC SCHOOL PUBLISHING & HAMPTON-BROWN

Sheron Long, Chairman

Samuel Gesumaria, Vice-Chairman

Alison Wagner, President and CEO

Susan Schaffrath, Executive Vice President, Product Development

Editorial: Fawn Bailey, Joseph Baron, Carl Benoit, Jennifer Cocson, Francis Downey, Richard Easby, Mary Clare Goller, Chris Jaeggi, Carol Kotlarczyk, Kathleen Lally, Henry Layne, Allison Lim, Taunya Nesin, Paul Osborn, Chris Siegel, Sara Turner, Lara Winegar, Barbara Wood.

Art, Design, and Production: Andrea Cockrum, Kim Cockrum, Adriana Cordero, Darius Detwiler, Alicia DiPiero, David Dumo, Jean Elam, Jeri Gibson, Shanin Glenn, Raymond Godfrey, Raymond Hoffmeyer, Rick Holcomb, Cynthia Lee, Anna Matras, Gordon McAlpin, Melina Meltzer, Rick Morrison, Christiana Overman, Andrea Pastrano-Tamez, Leonard Pierce, Cathy Revers, Stephanie Rice, Janet Sandbach, Susan Scheuer, Margaret Sidlosky, Jonni Stains, Shane Tackett, Andrea Thompson, Andrea Troxel, Ana Vela, Teri Wilson, Chaos Factory, Feldman and Associates, Inc., Tighe Publishing Services, Inc.

THE NATIONAL GEOGRAPHIC SOCIETY

John M. Fahey, Jr., President & Chief Executive Officer

Gilbert M. Grosvenor, Chairman of the Board

MANUFACTURING AND QUALITY MANAGEMENT,

The National Geographic Society

Christoper A. Liedel, Chief Financial Officer

George Bounelis, Vice President

National Geographic School Publishing

Hampton-Brown

P.O. Box 223220

Carmel, California 93922

www.NGSP.com

Printed in the USA.

ISBN: 978-0-7362-6225-5

10 11 12 13 14 15 16 17

10 9 8 7 6 5 4 3 2 1